Genetically Modified Crops:
Promises, Perils, and the Need for Public Policy

The impact of human activities on all peoples, as well as on nature, needs to be central in any forthcoming political discourse.

Societies are organisms that evolve through and with their members; they are not mechanisms to be assembled and disassembled at will. Yet the potential of doing just that exists to date on a global scale because of the characteristics of asynchronistic technologies. Thus the struggle to understand and steer the interaction between the bitsphere and the biosphere is the struggle for community in the broadest ecological context.

This is a collective endeavour that no group or conglomerate can do on its own. Most of our social and political institutions are both reluctant and ill-equipped to advance such tasks. Yet if sane and healthy communities are to grow and prevail, much more weight has to be placed on maintaining the non-negotiable ties of all people to the biosphere.

Ursula Franklin[1]

Quaker Institute for the Future Pamphlet Series

1— *Fueling our Future:* A Dialogue about Technology, Ethics, Public Policy, and Remedial Action by Ed Dreby and Keith Helmuth, Coordinators, with Judy Lumb, Editor, 2009

2— *How on Earth Do We Live Now? Natural Capital, Deep Ecology, and the Commons* by David Ciscel, Barbara Day, Keith Helmuth, Sandra Lewis, and Judy Lumb, 2011

3— *Genetically Modified Crops:* Promises, Perils, and the Need for Public Policy by Anne Mitchell, with Pinayur Rajagopal, Keith Helmuth, and Susan Holtz, 2011

Forthcoming:
4— *How Does Social Transformation Happen? A Guide to Values Development* by Leonard Joy

Quaker Institute for the Future Pamphlets aim to provide critical information and understanding born of careful discernment on social, economic, and ecological realities, inspired by the testimonies and values of the Religious Society of Friends (Quakers). We live in a time when social and ecological issues are converging toward catastrophic breakdown. Human adaptation to social, economic and planetary realities must be re-thought and re-designed. **Quaker Institute for the Future Pamphlets** are dedicated to this calling based on a spiritual and ethical commitment to "right relationship" with Earth's whole commonwealth of life.

Quaker Institute for the Future
<quakerinstitute.org>

Genetically Modified Crops:
Promises, Perils, and the Need for Public Policy

Anne Mitchell
with Pinayur Rajagopal
Keith Helmuth and Susan Holtz

—Quaker Institute for the Future Pamphlet 3—

Copyright © 2011 Quaker Institute for the Future

Published for Quaker Institute for the Future by *Producciones de la Hamaca*, Caye Caulker, Belize <producciones-hamaca.com>

ISBN: 978-976-8142-30-6

Genetically Modified Crops: Promises, Perils, and the Need for Public Policy is the third in a series of Quaker Institute for the Future Pamphlets:

Series ISBN: 978-976-8142-21-4

Chapters four and five were prepared as fact sheets by the Canadian Institute for Environmental Law and Policy for the Quaker Institute for the Future and are available online: chapter four <cfsc.quaker.ca/pages/documents/Helmuth1.pdf> and chapter five <cfsc.quaker.ca/pages/documents/QuakerBiotechFactsheet1.pdf>.

Producciones de la Hamaca is dedicated to:
—Celebration and documentation of Earth and all her inhabitants,
—Restoration and conservation of Earth's natural resources,
—Creative expression of the sacredness of Earth and Spirit.

Contents

List of Abbreviations   7

Preface – Anne Mitchell   8

Chapter One—Introduction: The Controversy Over Agricultural Biotechnology   – Anne Mitchell   11

   The Precautionary Principle: Cartagena Protocol on Biosafety   12

   Addressing Biotechnology in Public Policy   13

   Corporate Funding Changes Biological Research   14

   Raising Public Awareness, Influencing Public Policy   15

Chapter Two—Agricultural Biotechnology: A Review and Critique   – Pinayur Rajagopal   17

   The Advent of Biotechnology   18

   GM Crops Tolerant to Herbicides   20

   Consequences of GM Corn   21

   Effect of GM Crops on Agricultural Productivity   22

   International Assessment of Agricultural Knowledge, Science and Technology for Development   23

   The Organic Agriculture Alternative   24

   The Rise of the Agro-Ecology Alternative   26

Chapter Three—Why is Public Policy so Private?   – Pinayur Rajagopal   27

   Contesting Technology, Politicising Science, and Corporate Domination   28

   The World Food Situation, Public Information, and the Corporate Agenda   29

Safety Regulations and the Need for Research	30
The Economic Impact of Biotechnology	30
The Context of Risk Assessment, Mandated Science, and Public Policy Decision Making	32

Chapter Four—Genetic Engineering: Challenging the Worldview of Right Relationship
— Keith Helmuth 36

Genetic Rationality	38
Quaker Testimonies and the Biotech Orientation	38
Preparing to Engage Public Policy	41

Chapter Five—A Framework for Discernment and Action on Biotechnology Policy
— Susan Holtz 42

Spiritual Groundings	42
Assessing Ethical and Moral Implications	43
Not a Single Issue: Biotechnology's Different Applications	44
Recommendations for Better Managing of Biotechnology	47
Action Strategy Recommendations	49

Chapter Six—Quaker Contributions and Future Trends
— Anne Mitchell 51

Glossary	**54**
Endnotes	**57**
Bibliography	**59**
Contributors	**65**
Quaker Institute for the Future	**68**

List of Abbreviations

Bt = *Bacillus thuringiensis*

CCC = Canadian Council of Churches

CIELAP = Canadian Institute for Environmental Law and Policy

EED = *Evangelische Entwicklungs Dienst* (Church Development Service)

E.U. = European Union

GM = genetically modified

GMO = genetically modified organism

IAAKSTD = International Assessment of Agricultural Knowledge, Science and Technology for Development

IFOAM = International Federation of Organic Agricultural Movements

QIAP = Quaker International Affairs Programme,

RR = Roundup Ready

U.K. = United Kingdom

U.S. = United States

WCC = World Council of Churches

Preface

Transnational corporations are offering agricultural biotechnology and genetically modified foods as a solution to food scarcity and the alleviation of hunger in the world. This pamphlet considers some of the controversies surrounding agricultural biotechnology and critiques these claims. It also considers some of the ethical and policy development issues.

The purpose of this pamphlet is to engage concerned citizens in this discussion and to help contribute to the dialogue that society needs as it sets policy for agricultural biotechnology, **transgenic** crops, synthetic biology, nanotechnology and genetic modification. Research, development, and commercial application of these innovative technologies are rapidly evolving, but almost no public consultation is being facilitated and very little public policy oversight is being exercised.

I have been carrying this concern for over 20 years. I first raised it within my religious community, the Religious Society of Friends (Quakers), in 2000 on the patenting of the **oncomouse** in Canada. A group of Quakers came together to consider the patenting of life forms for commercialization. Our discernment brought us to the question; *"What is it in nature and human knowledge that we have the right to own?"* I then brought this concern to the Canadian Council of Churches (CCC) where the query was considered. The CCC agreed to intervene before the Supreme Court of Canada, arguing that the patenting of the oncomouse was a commodification of life. The Supreme Court decided in our favour by overturning the decision of a lower court to legally permit patenting the oncomouse. Canada is now the only G8 country that does not allow the patenting of higher life forms.

I next raised the question of agricultural biotechnology with Quakers in the Toronto area. A meeting was convened in May 2006 that brought in Quaker Institute for the Future. Eighteen people, including participants from other religious communities, gathered at Friends House in Toronto for a day-long meeting to begin a discernment process on how Quakers, and people of other faith groups, could be involved in the policy development process. A smaller group of Toronto Quakers continued to work with me on this concern. Our initiative came to be represented in the Biotechnology Reference Group that was established by the CCC to carry this concern forward within the larger religious community and at the level of national policy. I was the CCC representative to the World Council of Churches Consultation on Genetics and the New Biotechnologies held in Johannesburg South Africa in 2007.

I continued this project as a Board Member and Research Associate of Quaker Institute for the Future, taking it to three annual sessions of the Institute's Summer Research Seminar. The sharing and collaborative nature of these week-long seminars provided helpful insights and direction.

Although this research and pamphlet was developed within the context of the Religious Society of Friends, we know that a wide range of concerned people are involved with the issues we address. While the values and testimonies of Quakerism are implicit in our work, we hope that the ethical perspective we advance will resonate with folks from all groups as we work together to understand these issues and take action for the well-being of human communities and the whole commonwealth of life.

We do not claim to be providing a unified view from Quakers on the issue of agricultural biotechnology, but we are providing a considered view from several Quakers who have grappled with the issue over a considerable period of time. While we are all citizens of Canada, and are particularly concerned with the development of public policy in the Canadian context, the issues of biotechnology and public policy are global. The review and critique of agricultural biotechnology and the ethical and policy concerns covered in this pamphlet apply with equal cogency to the situation in the U.S. and around the world. Since the authors are Canadian, readers will find Canadian-British spellings consistently used. A Glossary of technical terms is provided and words included are in **bold** on their first usage.

I am grateful to my Oversight Committee of Toronto Friends Meeting, Friends, Rosemary Meier and Ellie Perkins who helped keep me focused. I am grateful to Pinayur Rajagopal, Keith Helmuth, and Susan Holtz, who contributed chapters to this pamphlet. Ellie Perkins, Shelley Tanenbaum and Natalie Caine's reading of the manuscript brought important points to our attention for which we are all grateful. Barbara Day generously provided her detailed editing expertise. We are grateful to the Quaker Institute for the Future (QIF) for agreeing to publish it in their *QIF Pamphlet* Series. As the *QIF Pamphlet* publishing agent, Judy Lumb served as the final editorial screen, did the design and layout work. We are grateful, as well, to Toronto Friends Meeting and Friends of Toronto Meeting who contributed the funds to have this pamphlet published. For all this important assistance we have an immense appreciation.

Anne Mitchell

Toronto, May 2011

CHAPTER ONE
Introduction: The Controversy over Agricultural Biotechnology
Anne Mitchell

Many people and organizations in countries around the world have concerns about **genetically modified (GM)** agriculture and particularly GM foods. These basic concerns include safety, health and environmental concerns: Is GM safe? What are the effects of GM on food and agricultural systems? Is GM decision-making democratic, open and transparent? Beyond these are additional concerns related to power and control, justice and equity.

Concerns around power and control focus on the rush to patent GM technologies resulting in corporate control and commercialization of any innovations at all costs. GM agriculture is just one example of technologies that are evolving fast. Others include synthetic biotechnology, geo-engineering, **nanotechnology**. The promoters of these technologies, both corporations and governments, are driving the agenda—including the research agenda—on the assumption that the benefits will be long-lasting and far-reaching for society through the generation of economic growth and wealth. Innovation and commercialization are seen as the means to be competitive in the global market place. Other factors, environmental, social justice, and equity, and long-term impacts on communities, are often downplayed.

In North America, decisions about GM are made by corporations and governments with little, if any, public input and without full and informed debate by our elected representatives. Little in the way of funding is allocated to consider the ethical, legal or social aspects of these new technologies, nor to ensure that the public is aware of the innovations that are occurring, their perceived benefits and any potential risks. Society—governments, the scientific community, and citizens—does not have the capacity to respond to the rapid innovations and commercialization of new products. These technologies are running amok. Risk assessment analysis needs to be broadened and the precautionary principle needs to be invoked.

The Precautionary Principle: Cartagena Protocol on Biosafety

The **precautionary principle** is an approach to decision-making in which new substances and processes are prohibited until tests deem them safe, rather than allowing usage until testing shows deleterious effects. In the case of GM foods, the U.S. regulatory agencies have taken the position that GM foods pose no new or unusual risks beyond that of foods without GM components.

One of the outcomes of the UN Conference on Environment and Development, also known as the "Earth Summit", held in Rio de Janeiro in June, 1992, was the adoption of the Rio Declaration on Environment and Development, which contains 27 principles to underpin sustainable development. Principle 15 states that, in order to protect the environment, the precautionary approach shall be widely applied by States according to their capabilities. Where there are threats of serious or irreversible damage, lack of full scientific certainty shall not be used as a reason for postponing cost-effective measures to prevent environmental degradation.

The required number of 50 countries ratified the Protocol by May 2003, so it became effective on 11 September 2003. Gatherings of signatories are called Meetings of Parties (MOP) and MOP VI is set to take place in Cartagena in 2012. Although 153 countries have ratified it thus far, the Protocol now hangs in the balance because the actual users of **genetic engineering** have not yet ratified it and are fighting against it, especially the U.S., Argentina and Canada.

The WTO judgment on the genetic engineering dispute, launched by the U.S. against E.U.'s ban on GM foods, disallows the defence of the E.U. as it had been based on the Cartagena Protocol as an international standard and the U.S. is not a signatory on the Cartagena Protocol. If the plaintiff in a WTO dispute is not a party to the international treaty the defendant bases their case upon, then it appears that the treaty cannot be used as a justification, even if it is the only standard relevant. Because of that ruling, the U.S., Argentina and Canada do not need to worry much about the Cartagena Protocol.[2]

Addressing Biotechnology in Public Policy

It is the responsibility of our governments to ensure adequate public policy is developed to protect the health and environments of citizens. Therefore, in addressing these technologies and their potential impact on society, governments need to consider questions such as: What's the purpose? Who benefits? Are there alternatives? Is the technology contributing to the common good? How do we define the common good? Governments are challenged to make appropriate decisions that consider aspects other than innovations and commercialization. Governments are challenged about how to make decisions and whom to engage in the conversation, particularly how to engage the public.

Innovative technology applications may be deemed "safe" by scientists, but there are many reasons for citizens to be concerned: lack of long-term studies; lack of consideration of societal or ethical dimensions; lack of understanding of impact on communities. Communities in the global south are particularly vulnerable. There is a view that new innovative GM agriculture will contribute to sustainability and food security. However, evidence is emerging that GM agriculture is destroying biodiversity, soil fertility, as well as communities and local, traditional agricultural practices. It is surprising to see the number of products that contain GM components in the food we buy in North America.[3] Our food labels do not say that these foods contain components that are genetically modified. In this pamphlet we cite scientific publications that show these products have not been tested for either short- or long-term effects. We also hear the voices of farmers and growers who are breaking through the layers of high expectations they have been flooded with, and are expressing what they are experiencing: high costs of seeds and fertilisers, effects on soil, and increasing appearance of pesticide-resisting weeds, some of which can be removed only manually.

Corporate Funding Changes Biological Research

When research is carried on in universities with public support it is assumed that the results of research will be published for sharing. Besides, the presence of other researchers in humanities and social sciences usually means the mingling of ideas and persons in discussions. Seminars by invited guest speakers are open to all. That is how science works—research results are published for the international scientific community to be able to share, critique, add, and advance knowledge. This was the way biological science was done until the advent of biotechnology. For the first time it was possible to create marketable biological products, so corporations became interested in biological research, which is very cost intensive. The rise of biotechnology occurred even as governments were reducing expenditures on funding for research. Science and medical research councils in Canada, the U.S. and the U.K. were cut back in part because of the pressure to reduce government, but also because defence expenditures were being increased. This reduction in funding for scientific research was presented to the public as the result of the need for reduction in public expenditures, even as the taxes for corporations were being reduced.

As governments reduce their research support and private corporations make inroads into the financing of biomedical research, the way of doing biomedical science changes. The corporations are not interested in the advancement of knowledge for the public good or knowledge for the sake of knowledge. They choose to support research that they can translate into marketable products; the patents arising from the research are the property of the corporation and not the universities. Research funded by corporations requires that the corporations have the ownership of the research. The funder has the rights to the results; they are not to be published in refereed scientific journals for the public to read. They become the proprietary intellectual property of the funder, and "intellectual property" was inserted into free trade agreements, so those rights are protected across borders. Subsequently, the results are converted into marketable products.

Thus the new biotechnology resulted in corporate funding of the research and that shift in funding has put a screen between what is publishable public information and what is privately owned by the funding corporations. Sometimes when the results in a university laboratory are promising, corporations either transfer the research to their own labs or set up institutes in the university and hire additional scientists who have appointments as adjunct university professors. The university scientists do the research and the corporation gets the output. The presence of such arrangements has supported many young research students during their early years, but their output is owned by the corporation.

Raising Public Awareness, Influencing Public Policy

There are other more urgent questions having to do with disharmonies already inherited by us. How can we discuss the harm already imbedded in the societal matrix? Do we have any processes to roll the dangers back? Do we even know what they are?

How did we get onto this slippery slope? Governments have vacated the policy arena in which they have the responsibility for policing corporate actions. The whole sector of overseeing regulatory bodies has been done away with in the interest of reducing government expenditures. The assumption, "not dangerous until proven otherwise", left free scope for companies to rapidly place products in the market and aggressively promote them in highly educated countries like Canada, the U.S. and the U.K. The process seems to be even more blatant in Asia, Africa and Latin America. Organizations in the U.K. are trying to remind the public of sustainable and ecologically responsible agriculture. Small farmers in Asia and Africa are trying to hold onto traditional practices of community-based agriculture. We have an obligation to help them understand that the scientific solutions recommended by biotechnology companies have not proven to be better than their own inherited practices, which are based on centuries of experience. However, is it too late to save our Canadian and American small farmers?

We hope this dissemination of information from articles published in refereed journals that speaks of the dangers and risks that lurk ahead for all societies is an important service Quakers can contribute through this pamphlet.

This pamphlet lays out some of the background, issues, and concerns about agricultural biotechnology. It also puts forward some suggestions for policy makers and citizens. It is intended for those who are interested in the science and technology and want to influence policy makers to ensure that transparent, effective policies that protect the common good are put in place. The pamphlet focuses on agricultural biotechnology, but the issues and concerns and the need for effective public policy in the public interest are similar across a broad range of innovative technologies.

We have organised this pamphlet in six chapters. In the second chapter Pinayur Rajagopal reviews and compares the promise and results of agricultural biotechnology. He draws on major research studies to show the measurable results of GM crop yields. He emphasises the importance of these research findings for a rational assessment of agricultural biotechnology, and provides access to this critical public information. In the third chapter Pinayur Rajagopal asks the question, "why is public policy so private?" He provides a window on risk assessment, science mandated in favour of industry, and public policy that favours private commercial interests. In Chapter four Keith Helmuth uses the framework of Quaker testimonies to explore the societal effects of corporate control of biotechnology research and its commercial applications. In Chapter five Susan Holtz provides a succinct guide on how to think about the complex field of biotechnology and outlines the range of concerns that are brought up when we think about its relationship to, and impact on, various factors of the common good. She provides a comprehensive guide for action strategies for both personal and citizen group engagement with public policy decision-making, as well as how to influence the behaviour of agri-industry corporations. In Chapter six Anne Mitchell shows how Quaker discernment and decision-making can bring an effective approach to citizen's engagement with biotech policy and the public interest. She includes a look at the trends in transgenic and nanotechnology research and development.

We hope that all readers who share our concerns will be able to apply this information and these action strategies to help develop an effective public voice on biotechnology. As these technologies evolve into the future, more and more support will be needed in defence of the public interest and the common good.

CHAPTER TWO
Agricultural Biotechnology: A Review and Critique
Pinayur Rajagopal

Even before biotechnology, there was an earlier panacea for feeding the world called the "Green Revolution". Norman Borlaug was said to be the "father" of the Green Revolution. When he died in 2009, we were reminded of the sweeping claims that have been made about its successes. It was claimed that Borlaug's "miracle wheat" doubled and tripled yields in a short period of time. Similar increases were soon achieved with maize, and with rice at the International Rice Research Institute in the Philippines. The success of the newly developed strains appeared limitless. They were introduced into several Asian countries and a number of African and South American countries. The spread of the miracle was heavily backed by U.S. Government support. The success story was propagated in mass media; and it was repeated in the media following the death of Borlaug.

Despite its obvious successes, however, the Green Revolution came under severe criticism during the 1970s for ecological and socio-economic reasons.[4] The main criticism directed against the Green Revolution was that high yields could only be obtained under certain optimum conditions—optimal irrigation; and the intensive use of fertilisers, chemical pesticides, and machinery, all of which required monoculture. Further, critics claimed that an important prerequisite for the Green Revolution mode of production was rich soil. Hybrid plants would otherwise be choked by weeds, which have adapted to the less favourable soils, and they could not survive the struggle against insect pests. Moreover, farmers living in problem regions were frequently too poor to be able to afford the expensive irrigation equipment and inordinate amounts of pesticides required.[5]

Sachs wrote about the misjudgements that characterised the first Green Revolution and mused about how to move towards a second Green Revolution.[6] He pointed toward organic agriculture, reliance on traditional methods, as well as drawing on centuries of local wisdom.[7]

"Today, as a consequence of technologies introduced by the green revolution, India loses six billion tons of topsoil every year. Ten million hectares of India's irrigated land is now waterlogged and saline. Pesticide poisoning has caused epidemics of cancers. Water tables are falling by twenty feet every year. The soil fertility and water resources that had been carefully managed for generations in the Punjab were wasted in a few short years of the Green Revolution's industrial mode abuses. If India's masses have avoided starvation, they have endured chronic and debilitating hunger and poverty."[8] India now exports food into the profit-driven, globalised, industrial food system, but 200 million mainly rural women and children still go to bed hungry.[9]

The Advent of Biotechnology

The field of biotechnology derives from the 1960s and early 1970s when molecular biology research developed the ability to isolate, replicate, and manipulate deoxyribonucleic acid (DNA), the molecule that carries genetic information. Realizing the immense potential of these techniques, the molecular biologists called for a moratorium on the research until the ethical and public concerns were fully discussed. Those discussions culminated in the Asilomar Conference in 1975 where guidelines were developed for containment and safe manipulation of GM bacteria and animal cells that were being used for basic research at that time.[10]

Because of the potential for biological products, biotechnology continued to develop under heavy financing by corporations, including those interested in agricultural products. Unfortunately, that research did not have the same public exposure and oversight as the original molecular biology research.

In order to introduce genes into the DNA of a plant cell, two barriers must be crossed, the outer cell wall and the nuclear membrane. Three methods have been used: 1) a micro-syringe has been used to inject the DNA directly into the nucleus viewed under a microscope; 2) the infection of the bacterium, *Agrobacterium tumefaciens,* can be used because it transfers some of its DNA into the genome of the plant cell; and 3) a **gene gun** fires pieces of DNA attached to heavy metals at plant cells at high velocities. Using these techniques, genetic engineering has created GM agricultural crops.

This agricultural application of biotechnology evolved rapidly and brought GM seeds for such crops as soybeans, maize (corn), canola, and cotton to the market beginning in 1996. These seeds have been marketed aggressively as governments and global seed companies push for new markets and higher sales. Often these markets are in developing countries and the sales pitch for this technology is accompanied by a promise of increasing yield, which is necessary to feed the expanding population. This technology is being pushed into use before farmers, consumers, and local policy makers can assess its potential negative consequences. However, there is an abundance of published information in refereed journals that contradicts the claims for GM crops. The purpose of this chapter is to review that information.

The U.S. Department of Agriculture reviewed the usage of GM seeds after 10 years of availability, indicating that they have been widely accepted by U.S. farmers. Seed companies submitted 11,600 applications for GM seeds and 10,700 were approved, a 92 percent approval rate. Nearly 5,000 of those were for corn and the rest were for soybeans, potatoes, and cotton. Most were for herbicide tolerance or insect resistance; others were for viral, drought, or fungal resistance. In 2005, GM herbicide-tolerant soybeans accounted for 87 percent of soybeans planted in the U.S., and GM herbicide-tolerant cotton accounted for 60 percent of total cotton acreage. GM Insect-resistant cotton was 52 percent of total cotton and GM insect-resistant corn was 35 percent of total corn. "U.S. consumers eat many products derived from these crops—including some cornmeal, oils, sugars, and other food products—largely unaware of their GM content."[11]

Economically, GM cotton and corn were associated with increased returns to the farms, but there was little impact of herbicide-tolerant soybeans on farm net income. The report found that although concern had been raised by consumers in the U.S., there was not much impact on the market for products containing GM-components. The U.S. does not require that labels indicate GM contents; non-GM foods are sometimes labeled "No Genetically Modified Contents", but those efforts have not made much impact on the market. However, in the E.U. and other countries, foods containing GM contents are labelled and consumer concerns have had a bigger impact, so there are very few foods with GM contents on their grocery shelves.[11]

GM Crops Tolerant to Herbicides

Weeds a problem? Apply more herbicides! This attitude is a good example of the industrial takeover of agriculture, a cautionary tale of technological failure. **Roundup Ready** (RR) crops are genetically engineered to be tolerant to the herbicide **glyphosate**, the active ingredient in **Monsanto**'s Roundup. Weeds that resist the herbicide glyphosate are adversely affecting RR crops on a large scale in the United States. According to Hammond,[12] for tens of thousands of American farmers, 2020 will be the year they revert to 1980s agricultural practices in order to control the rapidly spreading problem. While GM crops have stumbled in the past, the spread of glyphosate-resistant weeds is causing problems more severe and widespread than any GM crop failure recorded to date, negating the benefits of the herbicide. Farmers have no choice but to use more and more powerful chemicals on their crops. In many cases the weeds have proven impossible to control even with chemicals and increased tilling. The stalk of a mature Palmer weed can reach six inches in diameter, and can damage mechanical pickers, so it must be removed by manual pulling. In a seeming anachronism in the highly mechanised environment of American agriculture, farmers must now hire labourers to manually weed GM crops.[13]

Not only do the cost of GM seeds and pesticides cut into farmers' incomes, but they must also pay for human intervention with weed-pulling. This disaster is the direct result of the way corporate greed and short-sightedness has accelerated the chemical treadmill. The biotechnology industry has downplayed the severity and adverse impacts of glyphosate-resistant weeds, while working aggressively to come up with responses to the problem. Three of those suggested responses are: 1) subsidies for use of herbicides, 2) crops with enhanced resistance to glyphosate, and 3) herbicide-resistant **stacked traits** that include resistance to more herbicides, which requires still more herbicides!

In a review of independent studies, Pusztai and Bardocz[14] indicate that there is an urgent need to carry out systematic and direct studies, independent of the biotechnology industry, on the short- and long-term effects on animal and human health exposure to glyphosate in association with the appropriate GM crop. They conclude with an ominous caution: "With the presently cultivated huge areas of RR crops and the anticipated even-larger future extensions of this glyphosate-dependent GM crop technology, the potential danger to animal and human health needs to be dealt with in advance, and not if or when it occurs. If we consider that RR soybeans may themselves damage reproduction, a combination of similar, possibly synergistic effects of the GM crop with glyphosate could be a potential disaster in waiting."[15]

Consequences of GM Corn

In the same way that the agri-chemical industry places profit above the long-term viability of world agriculture, it is also subjecting both human and livestock populations to an uncontrolled metabolic experiment with regard to GM food crops. "For the first time a comparative analysis of blood and organ system data from trials with rats fed three main commercialised [strains of GM corn] ... which are present in food and feed in the world. ... Effects were mostly associated with the kidney and liver, the dietary detoxifying organs, although different between the three GMOs. Other effects were also noticed in the heart, adrenal glands, spleen and haematopoietic system. We conclude that these data highlight signs of hepato-renal toxicity, possibly due to the new pesticides specific to ... GM corn. In addition, unintended direct or indirect metabolic consequences of the genetic modification cannot be excluded."[16]

Effect of GM Crops on Agricultural Productivity

Gurian-Sherman of the Union of Concerned Scientists (UCS)[17] evaluated the overall effect genetic engineering has had on crop yields in relation to other agricultural technologies. Yield is the amount of a crop produced per unit of land over a specified period of time. The report reviewed two dozen academic studies of corn and soybeans, the two primary GM food and feed crops grown in the United States. Based on those studies, the UCS report concluded that the yields of GM herbicide-tolerant soybeans and herbicide-tolerant corn have not increased over conventional soybeans and corn. Insect-resistant corn that contains genes for insecticides from *Bacillus thuringiensis* (***Bt***) has only marginally improved yields. The increase in yields for both crops over the last 13 years was largely due to traditional breeding or improvements in agricultural practices.

The UCS report makes a critical distinction between intrinsic yield and operational yield, concepts that are often conflated by the industry and misunderstood by others. Intrinsic yield refers to a crop's ultimate production potential under the best possible conditions. Operational yield refers to production levels after losses to pests, drought and other environmental factors. The study reviewed the intrinsic and operational yield achievements of the three most common GM food and feed crops in the United States. The report found that all three GM crops—herbicide-tolerant soybeans, herbicide-tolerant corn, and *Bt* corn—have failed to increase intrinsic yields. Herbicide-tolerant soybeans and herbicide-tolerant corn also failed to increase operational yields, compared to conventional methods. The report found that *Bt* corn provided a marginal operational yield of 3 to 4 percent over typical conventional practices, over the 13 years. Since *Bt* corn became commercially available in 1996, its yield advantage averages out to a 0.2 to 0.3 percent yield increase per year. To put that figure in context, overall U.S. corn yields over the last several decades have annually varied approximately one percent, which is considerably more than the *Bt* traits have provided.

The UCS report comes at a time when spikes in food prices and localised shortages worldwide have prompted calls to boost agricultural productivity. Biotechnology companies maintain that genetic engineering is essential to meeting this goal. Monsanto, for example, is running advertising campaigns warning of an exploding world population and claiming that GM seeds significantly increase crop yields. The UCS report debunks that claim, concluding that genetic engineering is unlikely to play a significant role in increasing food production in the foreseeable future. The biotechnology industry has been promising better yields since the mid-1990s, but UCS report documents that the industry has been carrying out field trials with GM crops for 20 years without significant increases in yield.

An optimistically evoked international community repeatedly declares global commitment to poverty reduction. The very existence of absolute poverty constitutes an ethical imperative to apply new knowledge, including plant science, to alleviate limits on human potential. In the rush to apply new knowledge and technologies, there is little time to pause to consider how people in developing countries have been doing things—their technology—because of the assumption that our way of doing is superior to theirs. To combat hunger due to overpopulation and climate, we will need to increase agricultural productivity, but conventional plant breeding has outperformed genetic engineering.

International Assessment of Agricultural Knowledge, Science and Technology for Development

The International Assessment of Agricultural Knowledge, Science and Technology for Development (**IAAKSTD**) was initiated by the World Bank in 2002 as a multi-stakeholder group involving international agencies, governments, civil society, private sector, and scientific institutions. The IAAKSTD commissioned reports, the latest of which was published in 2009. Authored by 400 scientists from 60 countries, the report documents what has been accomplished so far.[18] Reservations about the report from the governments of Australia, Canada, and U.S. were included.[19]

Conclusions of this comprehensive assessment include:

- Farmers are the principal generators and stewards of crop genetic resources.
- The social and economic impacts of GM crops benefit large scale producers, but there is less evidence of positive impact for small producers in developing countries.
- The safety of GM foods and feed is controversial due to limited available data, particularly for long-term nutritional consumption and chronic exposure.
- Concerns are raised about implications of GM crops for biodiversity.
- Concerns are raised about the limited number of properly designed and independently peer-reviewed studies on human health.

"Taken together, these observations create concern about the adequacy of testing methodologies for commercial GM plants fuelling public scepticism and the possibility of lawsuits."[20]

The Organic Agriculture Alternative

In 1943 Sir Albert Howard, formerly director of the Institute of Plant Industry, Indore, and agricultural adviser to states in Central India and Rajputana, is considered to be the grandfather of the modern organic farming movement. He published *An Agricultural Testament* based on his years of patient observations of traditional farming in India. "Instead of breaking up the subject into fragments, and studying agriculture in piecemeal fashion by the analytical methods of science, appropriate only to the discovery of new facts, we must adopt a synthetic approach and look at the wheel of life as one great subject and not as if it were a patchwork of unrelated things."[21] Almost 70 years later, with the advent and adoption of GM crops succeeding the Green Revolution these words have returned to haunt us. The ongoing commercialisation of agriculture in India continues, with the U.S. extracting many pounds of flesh through trade agreements like the Indo-U.S. Knowledge Initiative in Agriculture, and U.S. Agency for International Development (USAID) and U.S. Department of Agriculture (USDA) investments in Indian agricultural universities. These agreements and investments have brought Indian agriculture under the full sway of GM crops controlled by Monsanto,[22] the 90% market leader. Monsanto is also on the Board of Directors of this initiative representing U.S. interests, along with other agri-industry giants.[23]

Organic and similar methods rely on a sophisticated scientific understanding of how a farm operates within an ecosystem, indeed, how the farm itself is an ecosystem of interconnected plants, insects, and other animals. Organic farming systems incorporate techniques like long crop rotations to control pests and leguminous cover crops or manure to add nutrients and build soil. A recent summary of studies on farming systems around the world found that such systems are often nearly as productive as current industrial agriculture in developed countries. The study demonstrates that the green and animal manures employed in organic agriculture can produce enough fixed nitrogen to support high crop yields. Where additional synthetic inputs are needed, other low-external-input methods are producing high yields with much reduced environmental impact. These highly productive methods are needed to produce enough food without converting uncultivated land into crop fields, removing forests that are important for biodiversity and slowing climate change. They build deep, rich soils that hold water, sequester carbon, and resist erosion. They don't poison the air, drinking water, and fisheries with excess fertilisers and toxic pesticides.[24]

The International Federation of Organic Agriculture Movements (IFOAM) has made the case for the benefits of letting countries follow traditional methods.[25] IFOAM is opposed to genetic engineering in agriculture in view of the unprecedented danger it represents for the entire biosphere and the particular economic and environmental risks it poses for organic producers. The reasons mentioned by IFOAM can be clustered into three groups: risks for human health and the environment, socio-ethical reasons, and incompatibility with the principles of sustainable agriculture. Kristiansen and Taii[26] present a summary of each of those arguments. Sir Albert Howard had it right 70 years ago.

The Rise of the Agro-Ecology Alternative

The Evangelische Entwicklungsdienst (EED) coordinated a joint project, over four years, in which 18 partners from all continents were involved. They looked into the question of whether or not genetic engineering is necessary to fight rural poverty and hunger. The results of their study produced a Reader in which reports from Maharashtra, Philippines, South Africa, Mexico, Tanzania, Bangladesh, Costa Rica and Brazil are included.[27] "Today in Asia and Latin America we are faced with the ecological consequences of intensive farming that paid little respect to the environment or society. The Green Revolution was concerned with introducing small farmers to modern industrial inputs, for example, the use of inorganic fertilisers, pesticides, and high-yield seeds. Yet, there is nothing to indicate that this approach in Africa functions any better now, than it did 30 years ago. How can we take account of the negative experiences in Asia? Even the new farming technologies are not making a major difference in this respect. The approach is false, because factors related to the local environment are ignored: humans, their culture, and society; soil types; climate; natural plant communities; the interplay of beneficial insects and pests; and other location-specific conditions. In addition, genetic engineering is full of risks and very expensive. Genetic engineering is no alternative to an agro-ecological approach, which is shaped by principles of diversity and improved with the involvement of local farmers. The agro-ecological, participatory approach not only promises better yields together with improved environmental conditions, but its distribution effect is more advantageous. It is of direct use to poor peasant farmers."[28]

If this agro-ecology approach to world food production and distribution can be steadily advanced, the costs and failures of industrial agricultural, with its GM crops, chemical arsenal, and inequitable wealth accumulation, may be forced to relinquish its grip on the future control of food.

CHAPTER THREE
Why is Public Policy so Private?
Pinayur Rajagopal

The previous chapter reviewed the results of assessments showing that GM crops have not increased crop yield and have created serious additional problems. Since this information is available to the public, why is it not being used by the public? As we prepare to act in the public interest on biotechnology policy concerns, we raise questions about science and technology, the economy, risk and precaution, and the common good. We ask about the integrity of science: is technology the servant or master of society? What is the economy for? What level of risk is acceptable within the guidance of the precautionary principle?

The book, *Right Relationship: Building A Whole Earth Economy*, provides an orientation to this kind of investigation.[29] It offers people from all philosophical and faith traditions, and all walks of life, an ethical guidance system based on "right relationship" between humanity and the whole Earth environment. This book was produced by the QIF Moral Economy Project which draws inspiration from the Quaker pioneers who planned the campaign that abolished the slave trade in 1789, and who then continued to fight for the abolition of slavery. John Bellers (17th C), John Woolman (18th C), and Kenneth Boulding (20th C) were Quaker economic thinkers who, along with many other Quaker social reformers, were important mentors guiding us toward a 21st Century understanding of the ethic of "right relationship". Over the years, Quakers have been given pioneering moral insights regarding the keeping of slaves, participation in war, incarceration in prisons, gender and sexual orientation equality, and, now "right relationship" with Earth's ecosystems. This is a new testimony. The book *Right Relationship* lights the way for people to grasp this new ethical grounding and be moved to take action without being inhibited by wondering, "Will what I do make a difference?" Joining with others to study, understand, and act in the public arena on biotechnology policy is the way to make a difference.

Contesting Technology, Politicising Science, and Corporate Domination

The intersection of technological change and human progress has often produced controversies. In the late 18th Century, Ned Ludd, and other small-scale weavers in England, smashed factory weaving machinery that was displacing cottage industry cloth production.[30] Factory owners and political leaders disparaged this action as an attack on "progress" and coined the characterization, "Luddite", for those who opposed technological change. But the action of the Luddites raised a critical social and ethical question: Whatever aggregate wealth production is projected by an economic cost-benefit ratio, there remains the ethically charged question of social disaggregation: whose benefit and whose costs? Is it right for the wealthy to pursue techniques that increase their wealth but have a disintegrating and harmful effect on the integrity of social and economic relations in the greater society? This question is still with us.

The genomic revolution resonates with previous contestations of technical change, but has raised genuinely new problems. The scope of this global dispute is reflected in titles of recent books: *Gene Wars; Pandora's Picnic Basket; Lords of the Harvest; Politics of Precaution; Seeds of Suicide.*[31] Science is the fulcrum on which this contentious politics rests. Science is an agnostic method for adjudicating truth claims. Applied in genomics, scientific truth is overwhelmed by a politicised science constructed to legitimise the strategies of corporations, government agencies, and politicians who receive the generous support of industry. Social movements and non-governmental organizations (NGOs) that apply the precautionary principle to genomic research and development are targeted for delegitimization by this politicised science. Science becomes less method than arena.

The World Food Situation, Public Information, and the Corporate Agenda

Historically, the causes of hunger and famine have been unrelated to food availability.[32] Eradicating global hunger is certainly a noble intention. "The U.S. agri-biotech industry and the scientific community had made a stand in defence of a mere 3,000 tons of GM rice; yet when the appeal's signers were informed that India had a food surplus of 65 million tons—non-genetically modified—yet a staggering population of some 300 million people went to bed hungry every night, they were not interested. Suddenly, their appetite for feeding the hungry evaporated and humanitarian intentions vanished into thin air when the fate of their precious GM rice was no longer the issue."[33]

What can we do in the West to help those in need through hard times? GM crops are not the answer. Instead "Make money available to buy food locally and distribute it. If that is not enough, you buy from neighbouring areas and distribute. If that is not enough, you get food in kind ... Priorities should be aimed at making that country competent in storing food and making food available in times of need."[34]

The apparent lack of information in the public, published realm about GM crops is something we have to counter. Well-intentioned people, scientists to boot, get drawn into joining appeals initiated by the agri-industry. The reader of such appeals may believe that when so many scientists have joined in the appeal there must be some truth to it, but unfortunately scientists who are experts in their chosen specialties, do not necessarily seek out conclusions of science in other areas before lending their support to appeals in areas outside their expertise. It is not difficult to understand why this can also happen in countries that are coaxed into accepting GM seeds.

In the U.S., people working in the agri-industry, in higher levels of government, and scientists working in institutes or laboratories funded by industries, commonly exchange positions. In February 2010 a delegation led by a U.S. Assistant Secretary of State arrived in New Delhi and was present there while the Indian Government was debating a moratorium on approving a GM vegetable, *Bt* brinjal, what we call "eggplant". The delegation included senior officials of Monsanto, who were former U.S. government officials, and scientists. They were addressing scientific audiences, political events, and seeking meetings with cabinet ministers. The presence of the U.S. Department of State officials in support of the GM-seed-producing corporation in India resembles the way the British used to push products on the colonies when India was a British colony.[35]

Safety Regulations and the Need for Research

In the European Union, the acceptance and regulation of GM crops/foods is based on the safety data that the biotech companies provide for the European Food and Safety Authority. The situation is worse in the U.S. where there is lax regulation. The commercialisation of GM crops/foods is based on the flawed concept of "substantial equivalence", that no credible evidence exists that GM crops harm human/animal health, so they are as safe as their substantially equivalent counterparts and need no safety testing. However, practically all recent reviews that have critically assessed the results of GM crops/food safety research data published in peer-reviewed journals have come to the conclusion that, at best, their safety has not yet been adequately established. At worst, the results of risk assessment studies, those carried out independently of the biotechnology industry, have raised important safety concerns that have not been properly settled.

The Economic Impact of Biotechnology

Wallace[36] reports on an investigation of the shaping of science, innovation and the economy in the U.K. and Europe. She begins with research funding decisions that are political decisions about how best to spend public money, which institutions to support, and what incentives to provide to researchers in academia and industry. Looking at the biological sciences, in the context of health and agriculture, the report describes how the idea of the "knowledge-based economy" has become a key driver of research investment in Europe and worldwide. However, the benefits of bioeconomy to the U.K. and E.U. were extremely limited.

- The net value of the bioeconomy worldwide was estimated to be zero or negative: with only two U.S. medical biotech companies, Amgen and Genentech, and one U.S. agricultural biotech company, Monsanto, making significant profits.

- Only two types of GM crops had been commercialised on any scale: insect-resistance and herbicide-tolerance. These crops are grown largely in North and South America for use in animal feed and subsidised industrial-scale biofuels.
- Concerns remain about environmental impacts, food safety, liability for contamination of non-GM crops and foods, and the extent of corporate control over seeds exercised through patents and licensing agreements.
- A number of new biotech drugs have been developed, but the U.K.'s only blockbuster biopharmaceuticals were discovered in the 1980s.
- Most new biotech spin-out companies from U.K. universities are not profitable and a net drain on the economy. They employ only 1,000 people in total.
- Genetic tests of multiple genetic factors are poorly predictive of common diseases and most adverse drug reactions; none are sufficiently predictive to meet medical screening criteria for use in the general population.

Wallace makes several recommendations on the need to re-assess what has been delivered by the major political and financial investments made in the bioeconomy over the past three decades, and to review whether current funding structures, institutions and review mechanisms are fit-for-purpose to deliver genuine solutions to the problems that we face. Review of the research funding system should lead to a major overhaul, which should include significant reforms to improve the scientific and technical advice available to everyone; reform the patents system; and re-structure funding institutions and systems of incentives for researchers. Objectives should include:

- More democratic decisions about research funding priorities and a more diverse research agenda;
- Greater accountability and scrutiny of major research investment decisions, including economic assessments and appraisals, scrutiny of scientific and technical assumptions, and active steps to prevent political "entrapment" in research agendas based on false assumptions and misleading claims;

- A role for public engagement in setting research questions and priorities: including consideration of a variety of alternative approaches to addressing problems, and greater democratic accountability for science policy decisions;
- More public engagement in research itself, which involves closer co-operation between universities, communities and civil society organisations;
- More funding for research which does not necessarily benefit large corporations but may deliver other benefits, such as public health research, and research into improving agro-ecological farming methods;
- Funding for "counter-expertise" and multi-disciplinary research which can identify long-term scientific uncertainties and regulatory gaps;
- Ensuring a thriving scientific culture that can analyse, critique, and develop the theoretical concepts that often underlie decision-making, and are key to developing new understandings;
- A commitment to take public opinions into account in decisions about science and innovation, including methods to ensure full consideration of the broader social, environmental, and economic issues associated with adopting particular approaches and technologies.

All of these are central to public policy making if policies and regulations are to benefit the public. Most of them are forgotten in the hurry to get the latest discoveries into practice. The U.S. Senate Finance Committee Chair, Charles Grealey, termed the relationship between regulators and regulated "too cozy", and this states the problem in a nutshell.[37]

The Context of Risk Assessment, Mandated Science, and Public Policy Decision Making

The precautionary principle obliges governments to assess environmental risks, warn potential victims of such risks, and behave in ways that prevent such risks. It puts the onus on the developer to show that an action is environmentally benign. A major limitation in the present regulatory framework is the lack of any clear sense of what constitutes environmental harm. Despite considerable debate, there are no guidelines that establish the magnitude of change that should trigger concern over ecosystem impacts. There are practical questions of how to assess long-term risks, and how much risk citizens are willing to accept. This ambiguity makes it even more difficult to develop an appropriate environmental risk assessment framework for agriculture.

The U.S. regulatory framework has taken the stance that transgenic crops do not pose any new or unusual risks.[38] Citizens, whose scientific literacy is generally fairly low, encounter biotechnology through a hyperbolic and distorted discourse. While previous Republican U.S. governments promoted GM crops, in Europe and Africa a powerful coalition of evangelicals and environmentalists opposed genetic engineering on theological and ecological grounds. The most notorious statement is perhaps that of the U.K.'s Prince Charles, who said that scientists working on GM foods had strayed into "realms that belong to God and God alone".[39]

Brunk *et al*[40] write about risk assessment as regulatory science. Using asbestos as an example, debates about risk often take this form: On one side are ordinary citizens, fearful that the asbestos in school walls and ceilings, or the radiation from overhanging power lines, will kill or maim their children. On the other side are officials of both government and industry who assure the citizens that there is no cause for alarm. Their sanguine view is bolstered by risk assessment experts whom they have called upon for an objective scientific opinion, based on the actual facts of the case. The alarmed citizens tend to be viewed as driven by well-intentioned, but scientifically uninformed, concern for the health of their children. They, in turn, may call upon their own group of scientific experts, whose conclusions lend support to their concern.

This raises an important question. Are risk debates disputes between those who accept the findings of science and those who do not? Between good science and bad science? Or is it possible that opposing assessments of risk, by scientific experts as well as ordinary citizens, reflect and are guided by dominant values held by assessors? An examination of one debate in Canada concludes that risk assessment is guided by dominant values held by the assessors.[32]

Even parties who claim to be using objective, value-neutral risk estimation methods and management criteria may find themselves in wide disagreement, as often happens among risk assessment experts. The criteria usually are not objective and value neutral, but are laden with normative and conceptual assumptions brought to bear on the process. So how objective is the science then?

Salter[41] uses the term "mandated science" to refer to the science that is used for purposes of making public policy. Science, here, includes studies commissioned by government officials and regulators to aid in their decision-making. It also includes scientific work originally produced in more conventional scientific settings. It becomes "mandated" when an individual study is evaluated in terms of the conclusions it can offer to policy makers about the merit of a particular regulation. The narrowing of science, as applicable to this particular proposed regulation, makes the whole process not a value-free enterprise. It makes it a value-laden enterprise.

Salter[42] shows three circles of activity: 1) public policy, regulation, law, and legal remedies; 2) values, norms, interests, conditions of use and trade and economic relationships; and 3) science. The intersection of the three is the sphere of mandated science. The intersection of values and norms of assessors, mandated science, and existing practices of law and regulation then confine how policy decisions are loaded in directions which are conducive to predetermined conclusions. If one adds risk-assessment on the top, a further layer of value assessments comes into play. When there is an assessment after a tragedy or calamity has occurred, the process may point to changing some existing rules. Other policies relate to anticipated risks that may arise in the application of existing practices, rules, regulations for GM crops. This is difficult in the U.S. where the regulatory framework has taken the stance that transgenic crops do not pose any new or unusual risks not faced by non-GM crops. When trade relations talks are being conducted, corporations might exert pressure on policy makers to make quick decisions so the corporation can synchronise preparations for a spectacular launch of a new GM product.

Corporations finance research in institutes and universities, and the results are their exclusive property. They spend money on persuasion and publicity. They reach decision makers, legislators, cabinet ministers, and those who are put on public policy bodies. Getting elected to public office is an expensive activity, and campaigns need funds. Often the funds come with strings attached. Even assuming that those in charge of making public policy go about making policy for the public good, the access to scientific truth can be hobbled by the terms of policy review. The research that is used for decision-making can be restricted to mandated science rather than all of science. The public is caught in the middle of all these pressures and publicities. Even the selective use of scientific expertise, as required by the regulations, narrows the frame of discussion. There is a lack of transparency in decision making. The public has to deal with all this while facing the realities of life: jobs, health, family support (including caring for members of older generations), all in the context of economies in crisis, global-warming, perceived terrorism threats, and wars into which their governments may have entered.

What do we, Quakers, have to offer in this situation? We do not have the funds to take the legislators, or all those who sit on bodies that make decisions that tilt in favour of GM crops, to court. What we can do is to assemble the best of available data, and set this in the matrix of Quaker concern and testimonies. There are many reasons for citizens to be concerned:

- Lack of long-term studies;
- Lack of consideration of societal or ethical dimensions;
- Lack of understanding of the impact on communities, particularly communities in the South where many farmers cultivate small plots of land and live under land ownership conditions that bind them to deliver to the owner of the land a pre-fixed amount at the end of each crop year;
- Claims of increased yield without making clear that there is the expectation of extensive use of fertilisers and pesticides, further exacerbated by the fact that GM seeds cannot be saved and replanted the following season.

CHAPTER FOUR
Genetic Engineering: Challenging the Worldview of Right Relationship
Keith Helmuth

Quakers have been reticent on theology, but articulate on worldview. Metaphysical speculation generally gets short shrift among members of the Religious Society of Friends (Quakers), but the worldview of right relationship is clearly focused in Quaker testimonies. The worldview of right relationship is experiential; it pays close attention to the conditions and processes of life, and matches decision making as closely as possible with the well-being of persons and places.

This approach served Quakers well as the culture of science emerged in the modernizing world, and its investigations greatly altered the common sense of reality. As scientific work advanced, the world has been shown to be composed not so much of fixed objects, as of relationships. True enough, discrete forms are plain to see, but the deep story of Earth and its commonwealth of life is now increasingly understood as process and relationship.

Because Quakers have understood continuing revelation to be the way things unfold, the new maps of reality drafted by science generally caused no difficulties in the minds of Friends. Quakers, in fact, became key figures in scientific work and helped create the scientific worldview. This worldview, until recently, rested on the assumption of a structurally dependable world, a world that may have its fluctuations and deviations but always returns to its ground rules of good order.

In recent times, however, certain avenues of scientific work have emerged that have given this worldview a real twist, and have caused a big hesitation, and sometimes a complete stop, in the minds of many folks, including some Friends. Biotechnology has changed the game with respect to the processes and relationships of life. The question of right relationship in this new complex of science and technology is highly problematic, and, in some respects, seems completely off the agenda.

Here are the main features that make biotechnology and bioengineering a special problem for the worldview of reverence for life and right relationship.

1) Genetic engineering has allowed scientists to isolate and scrutinise single genes, thereby making possible the alteration of genetic material and its transfer from one organism to another.

2) These technologies have made it possible to move genes and the information they express within and across species.

3) Not only are these technologies transforming fruits, vegetables, and livestock, they are now poised to alter the human species in a variety of ways.

4) Industrialised food and pharmaceutical corporations are systematically altering crop seeds and plant-derived medicines in order to claim them as inventions and place them under international patent and trade regulation protection. This makes it illegal, in many cases, for farmers and indigenous peoples to save and plant seeds or prepare medicines that are related, even remotely, to the biotechnology products.

5) These technologies, from development through deployment, are fully immersed in the swift currents of capital-driven economics.

6) Capital-driven economics is marked by neo-liberal free trade accords that favour large corporations, and the accumulation of wealth that is widening the divide between rich and poor, both within and between nations.

If we take human solidarity as our moral compass, and right relationship as the map of Friends testimonies, we can explore a Quaker approach to these new technologies. Before going further, however, something of the worldview of genetic research should be put into this picture. Of all the conceptual arrangements through which rationality can be focused, one, in particular, carries the day in biotechnology and the culture that surrounds it—genetic rationality.

Genetic Rationality

Genetic rationality[43] came into effect when it was first understood that genes carry traits, that traits are inherited, and that selective reproduction can emphasise or repress traits. If this was, indeed, the key to understanding continuity and change in life process, the rational mind had now found the biotic holy grail. With the development of genetic engineering, all other factors of significance in the history of life fell to a subordinate level, and genetic rationality moved into a command and control position over virtually the whole range of biotic expression. This convergence of an unprecedented and extraordinarily powerful technology with the bias of rationality has created an almost irresistible momentum for open-ended manipulation of biotic process and form. The bias of rationality is to be always on the hunt for the *one best way* in whatever field is being investigated. In genetic engineering, rationality has found a royal road, a *one best way* to investigate and change the biotic world.

What can it mean to apply Quaker testimonies to biotechnology and the culture of genetic rationality? Is the worldview of right relationship in any way commensurate with the power of this research? How does the ethic of right relationship understand and address this new engine of science, technology, and wealth creation?

Quaker Testimonies and the Biotech Orientation

A brief review here of Quaker testimonies, along with some contrasting characteristics of genetic engineering and its commercial development, will help create a platform from which further scrutiny can be launched. Quaker testimonies are not rules for living, but expressions of living witness that have emerged from Quaker experience. They are generally characterised as simplicity, peace, integrity, equality, and community.

Simplicity: Simplicity is, in large part, about focusing on relationships and processes that are fundamental to a well-balanced life. In practice, this can be fairly complex. A well-balanced life may be composed of many elements, but if these elements intersect with a high degree of right relationship, a kind of ordering principle—a kind of higher simplicity—emerges in our sense of guidance and well-being.

Biotechnology has a very different orientation. It is not interested in achieving balanced functioning within natural and social systems. Natural and social systems are often the problem it seeks to overcome. Biotechnology is aimed at unbalancing natural and social systems in favour of controlled, selective benefit for commercialisation and capital accumulation.

Peace: The Quaker peace testimony manifests in both personal life and in larger social forms. Here too, the ethic of right relationship serves the full spectrum of peace concerns. The domain of peace includes nonviolent living, conflict prevention and resolution, and reducing the causes of conflict, violence and war.

It is well known that war and preparation for war stimulates scientific research and technology development. Biotechnology is no exception. It is firmly ensconced in the military saddle.[44] The U.S. national security establishment is deeply involved with biotechnology. The biotechnology we hear about is mostly aimed at health improvement and yield per acre of food crops. But, apparently, this is just the opening act. The full story of biotechnology aggression and domination is yet to come in such developments as genetically engineered "war fighters", food crop control, population management, and even cognitive and emotional manipulation.[45]

Equality: The testimony on equality is best thought of as the ethic of equity. Equity means a fair share, a valued status, the prospect of a fulfilling and productive life. It means recognition and respect, and the life circumstances that draw out and support human dignity.

Biotechnology's relationship to equity is complex and increasingly problematic. The promise of genetic engineering in agriculture, having run into a myriad of unforeseen problems and unintended consequences, has mostly stalled. Through inequitable trade agreements and quasi-legal regulations, agri-industry and pharmaceutical giants are systematically enclosing the genetic commons, turning germ plasma into a commodity over which they then have exclusive control. Because biotechnology has developed mainly within the domain of capital-driven economics, its products and services are, and will be, available only to those who can pay for them at a level that advances capital accumulation for the corporate investors. Thus the already rich get richer and inequality advances.

Integrity: The testimony on integrity is a linchpin testimony. It vitalises and validates all the rest. At the first level it encompasses truthfulness and ethical consistency. In a widening perspective it includes devotion to right relationship and the high valuing of direct experience in the formation of knowledge and judgment.

Biotechnology, on the other hand, is committed, front and centre, to the manipulation of integrity. It works for the enclosure, monopolistic control, and commercialization of integral biotic components, and for their excavation and transplantation into now redesigned organisms that will yield market value. Here we meet the full force of the biotechnology industry. To *the one best way* of genetic rationality it has added *the one best way* of market rationality.[46] This is a formidable cultural alliance that, in effect, has become a new religion,[47] a new faith on how humanity, or at least part of it, should proceed into the future.

Community: Largely because Friends have had an enduring concern for right relationship, and because Friends have a well-tended tradition of collaborative discernment in decision-making, the soul of community has been kept alive in Quakerism. Community has thus become a special witness and testimony of experience, a witness for right relationship, and a testimony about communion.

In contrast, biotechnology has no particular interest in community. The industry is focused on the individual components of organisms and on individual organisms. From the office towers of Monsanto to the Wister Institute at the University of Pennsylvania, biotechnology, whether dealing with a rice plant or a research subject in a therapy trial, has a consistent point of view—the state of the individual organism and the opportunity to modify it. No thoughts of community trouble biotechnology's focus on the individual. This is a dangerous blind spot in so powerful an industry. With respect to the integral reality of human experience, community is the key to resilience and a hopeful future, not individual organisms and genetic manipulation.

Perhaps here, most of all, Quaker discernment[49] and testimonies are called to ask: What about community? What about equity? What about solidarity? What about right relationship? What about the commonwealth of life? What good is a high yield crop if an increasing number of people can't afford the price? What good is almost perfect health or enhanced cognitive powers if you don't have a functional community and a world at peace?

Preparing to Engage Public Policy

From the standpoint of Quaker testimonies, three questions about biotechnology and other powerful new technologies come into view: 1) How can their benefits be developed and applied in an equitable way, a way that serves the common good? 2) How can damaging and potentially disastrous consequences to ecosystems and social systems be foreseen and forestalled? 3) Are there zones of organic process and ecological relationships that should be ethically off limits for genetic engineering? Addressing these questions requires us to understand the convergence of moral, ecological, economic, and political realities that takes place in the field of biotechnology. With such understanding in place, we can then seek out and help create the opportunities for public engagement with public policy on the development and regulation of biotechnology.

We can speak with confidence on the ethics of right relationship in science, politics, and economics, and on the ecological integrity of the whole commonwealth of life. There is a mounting wave of moral energy looking for effective focus on the convergence of justice, peace, and ecological integrity. Quaker testimonies, which are a cluster of values clearly shared by many people worldwide, are a prime site for hosting this convergence. Thus can Quaker testimonies enter fully into the life of our society and the world on this critical issue at this critical time.

CHAPTER FIVE
A Framework for Discernment and Action on Biotechnology Policy
Susan Holtz

There are three basic questions that emerge when Quakers consider the various topics and issues in the large field of biotechnology:

- What fundamental moral and spiritual groundings does biotechnology touch on and do Quakers need to be in agreement on these matters?

- How should we go about assessing biotechnology's ethical and moral implications?

- What could we do about biotechnology's direction and management if we are moved to take action, collectively and individually?

Spiritual Groundings

Topics here involve perspectives on science, cosmology, and on life itself, and our human role in Creation, or the world, depending on your theology. For example, many might believe that it is inappropriate to manipulate certain natural aspects of biology, but discussing which aspects, and why, would likely raise deep questions about underlying beliefs. Moral fundamentals also include Quakerism's core testimonies and biotechnology's implications in light of those testimonies. Keith Helmuth's discussion in the previous chapter provides a detailed and thoughtful commentary on this latter subject.

Clearly these are important matters for members of a faith community, but do we need to seek unity about them? I don't think so. One of Quakerism's characteristics is an avoidance of dogma and creedal rigidity in favour of openness to inquiry and lived experience. The resulting diversity is a source of spiritual nourishment that most Quakers cherish. This is especially so regarding the big philosophical and theological questions that biotechnology raises. And while some Quakers may feel that certain issues have such significant spiritual and moral implications that these become the heart of the matter for them, the wellspring for practical concern need not be the same for everyone. There is still much room for consensus about specific issues and for taking collective action to address them.

Assessing Ethical and Moral Implications

When it comes to assessing practical positions related to ethics, morality, and appropriate public policy, we need to be much more concrete about what the issues are than when we're engaged in a broad philosophical discussion. In particular, we have to identify the specific applications of biotechnology on which we're focusing. Each one poses distinct questions that may suggest quite different answers in light of different circumstances.

Ethics is not exactly the same thing as morality. Morality is about characterizing the nature of the Good and Wrongness, the Bad, or Evil, with or without the capital letters, depending on your philosophical bent. In contrast, ethical issues are about right action within a very specific social and economic context. The context, more than the action, is the crucial factor. For instance, the ethics, as opposed to the morality, of a sexual relationship between consenting adults depends very much on the institutionalised power relationships between the individuals: whether they are doctor-patient, student-instructor, or boss-subordinate, for example.

Many of the issues related to biotechnology are as much, and sometimes more, concerned with ethics than morality, though there are exceptions, as discussed in the next section. Particular ethical and moral questions are often similar to those about other technologies and public projects. A pragmatic approach often focuses on benefits and risks. We need to ask and answer questions like these:

- What exactly are the benefits?
- Who in particular benefits?
- Is that arrangement fair to individuals or individual actors, like corporations?
- Is that situation consistent with a social justice vision for society?
- What exactly are the risks?
- Are the risks equitably borne among all those affected?
- Are the risks acceptable or not, and on what basis?
- Do the risks outweigh the benefits, and why or why not?

Good information, intellectual and ethical integrity, and sound judgment matter here. There must be an understanding of all the facts, including what is known and not known, and careful discernment in assessing the relative weights of risks and benefits. Being able to explain clearly the reasoning behind a decision is also important.

Not A Single Issue: Biotechnology's Different Applications

As noted earlier, perspectives on these questions may vary, depending on what application of biotechnology is under discussion. I would suggest that it is useful to classify the hundreds of different biotech applications into six different categories that bring up different sets of issues and different possible responses. These six groupings of biotechnology applications follow.

Basic research

Some avenues of research raise questions that are really more about spiritual or moral unease concerning the implications and directions of this research than about specific ethical issues. An example is the synthesizing from scratch of the entire genome of the bacterium *Mycoplasma genitalium*.[49] This is a major step leading toward the creation of synthetic organisms. Some may believe that such research is morally or spiritually wrong, regardless of any practical risks or benefits. Others may disagree.

Medical research and reproductive technology

This generally involves issues related to individual privacy, cloning of humans and other animals for use, and decisions about "designing" certain human characteristics. Sometimes there are definite moral dimensions to these questions. The Canadian Supreme Court decision disallowing the patenting of the "oncomouse", for instance, was argued in part on the felt sense of inappropriateness about owning patent rights to a higher life form. Much conflict centres on common types of ethical dilemmas within historical and evolving medical practice, where rights, risks and benefits for individuals interact with public policy. Another dimension of these applications is the use of traditional knowledge about medicinal plants and remedies and their genetic manipulation for corporate profit. These issues mainly concern public policy matters, such as, patent rights and compensation for the appropriation of traditional knowledge.

"Indoor" biotechnology applications

There is a wide range of very different applications here. In a field like renewable energy, for instance, researchers are investigating how micro-organisms might be genetically modified to produce renewable transportation fuels, such as ethanol, biodiesel, or even gasoline from biological source materials like waste wood or algae. Another huge area of activity is human health and medicine, with research involving genomics, gene therapy, and diagnostics, among other topics. A more problematic example is the use of genetically modified plants to produce pharmaceutical products or industrial chemicals. Such diverse applications are grouped together in this category if they take place inside secure facilities, including factories, greenhouses, or laboratories, with the intention of strictly containing the modified organisms. This is because the risks they pose are clearly ecological, directly related to human or animal health, or perhaps to personal privacy. The critical questions for this category are whether strict containment is possible and the severity of the consequences if it is breached. Environmental and health benefits from these applications are generally considered significant, though with various reservations.

"Outdoor" biotechnology applications

These include applications involving GM seeds, crops, and trees for commercial use in agriculture and forestry, outdoors in the open environment. Such applications have been highly contentious, involving ethical objections related to ecological risks, unknowns regarding health risks, and social justice issues about the increasing concentration of corporate control in agriculture, along with negative economic impacts on small farmers in the developing world and in the organic agriculture sector everywhere. The claims of economic benefits, along with the suggested need for GM crops to provide higher yields and reduced environmental damage from tillage and pests, are also very contentious.

Military applications

Quakers and members of other "peace churches" will obviously have objections to such applications. If there are risks to civilians and the environment, others may also oppose developments in this area.

Hybrid bio-nanotechnology

Hybrid bio-nanotechonology is sometimes referred to as next-generation nanotechnology. Biotechnology does not actually include nanotechnology, which is about the manipulation of materials at the molecular scale. Nanotechnology poses its own set of risks, since materials at this extraordinarily small scale have novel physical, optical, electrical, and other characteristics. Little is known about the toxicity and other biological and ecological effects of nano-scale materials, but because of their various special properties, such as electrical conductivity or great tensile strength, they are already being used in upwards of 600 commercial products.

However, next-generation nanotechnology is about marrying genetic engineering with nano-scale materials in order to make self-assembling products, such as solar arrays, battery components or even batteries. There are many concerns about ecological, worker health, and other risks, mainly because regulation of nanotechnology is not in place and many would argue that regulation of biotechnology is inadequate.

In the summer of 2007, Environment Canada posted notification that nanomaterials having novel molecular structures would be subject to the Canadian Environmental Protection Act regulations, an important step in the right direction. Nevertheless, that still excludes from regulatory oversight many nanomaterials now in use, though there is ongoing discussion about further developments in nanotechnology regulation and policy. So far, putting in place a regime for managing and regulating hybrid bio-nanotechnology has received virtually no public or political attention at all.

Recommendations for Better Managing Biotechnology

Managing technology is never a completely successful enterprise. Still, public policy has a number of institutional means to try to do this in order to achieve social or economic goals. What is striking about biotechnology is how narrowly these goals have been defined—commercial success, primarily—and how limited public input has been into policy decisions about its goals, research, financial support, and regulation.

In Canada there are six key aspects of biotechnology policy that are clearly problematic and merit institutional change. In the United States and many other jurisdictions, these areas are also likely to be among the weakest and least adequate for appropriate management of this technology. They are as follows:

Public research funding

Formal avenues for public input and review are required.

Policy review and advice

Formal avenues for public input and adequate support for public participation are lacking and should be developed.

Labeling

Currently and for many years past, the Canadian government has refused to require labeling of food and other products involving **genetically modified organisms (GMOs)**. This decision deprives citizens of the ability to make informed personal choices about these products. Especially in a contentious political climate regarding GMOs, information and transparency via labeling allows market signals to influence choices made by food producers and retailers, and directly rewards those giving customers what they want to buy.

Regulatory regime

The present regulatory regime is a confusing patchwork of existing legislation that may or may not be adequate to manage the specific challenges of biotechnology. It requires a thorough review, and possibly major change to consolidate, streamline, and make it more transparent and effective.

Liability

Currently, there is no general legislative direction in Canada concerning civil liability for biotechnology. For those who might be affected—for example, an organic grower whose crop is contaminated by neighbouring genetically modified crops and who loses organic certification because of this—the only choice is a civil suit with the burden of proof on the complainant. Unlike some other countries, Canada has no legislation apportioning responsibility and costs to either the neighbouring farmer or the manufacturer. The result is legal uncertainty and financial exposure for those adversely affected. As well, farmers with GMO-contaminated crops are liable to civil action against themselves for patent infringement, despite the fact that it is now known that accidental genetic contamination of nearby plants from GMOs can and does take place.

World Trade Organization (WTO) Rules and Trade-related Issues

Current international trade rules constrain national initiatives and international treaties from broadly employing some approaches to environmental legislation, and related economic and trade sanctions. This is not entirely unreasonable, as some countries have used "environmental" legislation as a protectionist mechanism for unfairly restricting international trade. However, right now, trade measures are one of the key points of conflict both within the E.U. and between the E.U. and North America regarding the acceptability of GM food crops, with an individual country's ability to restrict GM crops at question. Many people would like to see a new round of WTO negotiations aimed at finding a more acceptable approach to environmental matters in both international treaties and in environment-related trade disputes. Most would agree with rules that disallow arbitrary "environmental" trade restrictions, but permit countries more latitude in their environmental restrictions involving legitimate environmental goals and considerations, especially in disputes where there is scientific uncertainty and the precautionary principle is invoked.

There are a number of options that concerned citizens can use to try to alter public policy and institutions related to biotechnology, and Quakers are familiar with most of them. The following is a listing of well-known strategies that are relevant to biotechnology issues:

- Raising public and political awareness through media campaigns and electronic media, including blogs, e-mail, chat rooms and other electronic venues;
- Energizing NGOs, including Quaker and other church organizations, to campaign on specific biotechnology issues;
- Directly exerting public pressure by organizing public events and through personal interaction with politicians, bureaucrats, and other influential organizations and individuals;
- Personal involvement in electoral politics;
- Initiating legal action where justified;
- Making a public commitment, collectively or individually, to purchase only or mainly organically produced food for a specified period while publicizing and explaining that action; and
- Organizing well-thought-out consumer boycotts of specific GM products and/or of companies opposing GM labeling.

However, it should be stressed that effective consumer boycotts must have a concrete objective that the target of the boycott can accomplish by its own actions, or at least very directly affect. Organizers of the boycott must be in a position to determine when that objective is reached and be able to then publicly call off the boycott while pointing to its success.

Action Strategy Recommendations

Quakers, and all others who share these concerns about biotechnology, could take action in various ways on all of the areas for institutional change—and I know that a number of people are indeed doing just that. But if I had to pick only a couple of priorities, the strategic actions I would recommend would be to:

1) concentrate on public awareness;

2) concentrate on energizing NGOs, including Quaker and other church organizations; and

3) concentrate on organizing a major consumer boycott coupled with organized, collective, well-publicised commitments to buying organic products. Support for low-income people in this endeavor should be part of the effort since one of the claims sometimes made for biotechnology is that it lowers consumer prices. This aspect should receive more attention.

Why these priorities? Unlike, say, climate change or air pollution, there is little public awareness or knowledge about these issues. Significant political changes in institutions are unlikely unless there is some perceived public pressure and support for such action. A boycott to mandate GM labeling, however, taps into existing consumer pressure for more information and transparency about what is in food and other consumer products. And buying organic addresses a range of ecological problems that many people are concerned about already, as well as having the effect of avoiding GM food products.

This is not to say that other institutional changes don't matter, only that they will require the collaborative work of many people and organizations, not religious groups alone. Specialist expertise and adequate resources are needed to initiate legal action, and a massive international push will be required to gain any traction on a WTO round of negotiations on the environment. Nevertheless, there are plenty of legal and trade experts who see this as a necessary institutional evolution. And there are NGOs who similarly see public input and regulatory reform of oversight of biotechnology as vitally important. Significant institutional change will require a strong effort to reach out to others to make common cause on these issues, and Quakers, along with others, could certainly help in providing some committed and knowledgeable leadership in doing that.

Chapter Six
Quaker Contributions and Future Trends
Anne Mitchell

The quest for responsible public policy must confront and negotiate a range of issues and a panorama of forces working for different visions of the future. Those devoted to the science of ecological understanding, the ethics of right relationship,[50] and the ancient bond between human communities and well-cared-for land, can take heart in the reports that organic farming is now the only segment of agriculture that is expanding. The longer term field results of agricultural biotechnology are now coming in, as are comparative results from organic agriculture. In the age of peak oil and climate change, eco-agriculture is gaining favour with farmers, because it makes sense both scientifically and economically. It is also gaining favour with the public. The quest for responsible public policy on food resources is not just a matter of assessing the risk factors of industrial agri-business, but it is also about advancing local sustainable food systems based on the agro-ecology approach. The current trends are well described by Geoff Tansey in *The Future Control of Food.* [51]

At the beginning of this pamphlet we raised the question of how best to publicise the kind of information gathered here. We can draw inspiration from the English Quaker Economist, John Bellers,[52] who, in the late 17th and early 18th centuries, was indefatigable in his efforts to induce political leadership to take up programmes of social and economic reform. He repeatedly petitioned both Houses of Parliament, providing all members with copies of his proposals. He wrote again and again to the Yearly Meeting, the Quarterly

Meetings, and the Monthly Meetings of the Religious Society of Friends urging them to act in concert on specific programmes that would alleviate poverty and benefit the nation as a whole. John Bellers helped pioneer the Quaker ethic that moved religious faith into the great work of social equity and human betterment, which has inspired us to provide the information and guidance collected here. It is our hope that Quakers, and many others who share these concerns, will find it helpful for advancing a public policy witness on biotechnology and the integrity of the commonwealth of life.

In addition to the testimonies and values detailed in Chapter four, one of the practices Quakers can bring to deal with the conundrums of GM agriculture and other innovative technologies is the old Friends' practice of "scrupling". Scrupling has been used in the past by Quakers to consider issues such as war and slavery. More recently Toronto Quakers organized a Scrupling Session on the erosion of democracy in Canada. Scrupling is not a debate, an argument or a panel discussion. Rather it is a process where one searches one's conscience about what is the right way forward and shares this with those gathered, not to reach consensus or unity, but to hear what others are saying. Scrupling could be used to consider the issue of risk in relation to the new bio- and nano-technologies. Following this practice of bringing conscience, moral awareness, and ethical direction to bear on the issues raised by these technologies, Quakers can work with those in the broader faith and secular communities to expand and develop public policy guided by the common good.

Quakers also have a long and well-practiced tradition of collective decision-making in which an open process of discernment and collaboration works toward unity on whatever matter is under consideration. This process is not the same as consensus, which is often reached through negotiation and compromise. The Quaker decision-making process is based on the experience that when collaborative discernment is practiced in openness and truth seeking, new light is often revealed that helps create better decisions than any of the participants may have had in mind going into the process. When matters are contested and the issues needing resolution are important, this kind of decision-making process can result in bringing diverse views into unity on the right way forward. If the parties and interests involved are committed to the common good, Quaker decision-making practice may be helpful in facilitating public engagement on important issues of public policy, such as agricultural biotechnology.

In entering the dialogue on biotechnology, public policy, and the common good, we should ask the question: Do we proceed according to an ecologically informed worldview of *right relationship*, or do we follow a *reductionist* worldview into an *instrumentalist* manipulation of life and relationships? Central to this process of discernment is the question of the commodification and control of life. It is no exaggeration to say these new technologies have the potential to change how we think about life. They may alter our understanding of what it means to be human. They have the potential to enclose community and societal development within a set of power relationships focused only on market transactions and the accumulation of wealth by an elite group.

These new, innovative technologies include synthetic biology; transgenic manipulation, bioengineering and cloning, nanotechnology, and geo-engineering. How will these technologies evolve? Will there be opportunity to develop public policy that defines and considers the common good? How will the great number of people deeply concerned about these technologies respond to issues of food and human security, control and access to seeds, and military applications? Can we influence those who hand out research grants to include a requirement that a percentage of each grant has to be spent on actively researching and calculating the full range of potential consequences, and make sure that the public is informed on the risks as well as the potential benefits? These technologies now influence how food is grown, how food products are made and marketed, and how other resources are utilized. They also influence decisions on the beginning and end of life, and on both community and environmental health. These are major societal concerns. Quakers, and all others devoted to the well-being of their communities and their society, need to be involved in a quest for responsible public policy on these new technologies. We hope this pamphlet will serve as a useful tool toward this end.

Glossary

Bt = *Bacillus thuringiensis*, a soil bacterium that produces insecticides, the genes for which have been inserted to enable corn to resist several kinds of insects.

Cartagena Protocol on Biosafety = an international agreement on biosafety, a supplement to the Convention on Biological Diversity that invokes the precautionary principle.

Gene gun = one of three basic processes of genetic engineering involves a gene gun, an appropriate name for such a violent technology. Microscopic pellets of gold or tungsten are coated with DNA from one or several organisms. The gene gun fires these pellets into another organism's embryo cell at high velocities (about 1400 feet per second or 1000 mph). Some of the DNA from the pellets enters the nuclei of the host cells, randomly inserting foreign DNA into the host's genome.

Genetic engineering = the practical application of basic research in molecular biology that allows isolation of genes, their replication and introduction into other organisms.

Genetic modification = inserting genes in the form of DNA into the genome of the recipient organism. Besides the gene gun described above, two other methods have been used. A microsyringe might be used to insert DNA directly into the nucleus of a plant cell. Or the infectious process of a virus or bacterium can be used to insert the DNA. In the case of plants, *Agrobacterium* has been used.

Genetically modified (GM) = changed by the application of genetic engineering. At the outset of the biotech era in the 1970s, "genetic engineering" was the widely accepted term to describe the brand new technology of gene splicing and recombinant DNA. In 1988 the European Commission adopted the term "genetically modified" in its first policy directive on the release of engineered organisms. In this pamphlet we have used this term abbreviated "GM".

Genetically modified organism (GMO) = an organism—plant, animal, or microbe—that has a modified genome with additional genes being added or genes being deleted or disabled, called "knock-out" organisms.

Glyphosate = the active ingredient in the herbicide Roundup produced by Monsanto.

Golden rice = a rice seed into which the gene for pro-vitamin A has been inserted. Pro-vitamin A is a biochemical precursor that is metabolized to vitamin A only in the presence of sufficient body fat. Vitamin A deficiency is supposed to be remedied by eating golden rice. However, a simple calculation based on the recommended daily allowance of vitamin A shows that an adult would have to eat at least 12 times the normal intake of 300 grams of rice to get the daily recommended amount of pro-vitamin A from golden rice.[53]

IAAKSTD = International Assessment of Agricultural Knowledge, Science and Technology for Development which was set up in 2002 for the purpose of assessing the effectiveness and impact of agricultural technics. The 2009 assessment report was authored by 400 scientists from 60 countries and published in seven volumes. The Synthesis Report (2009, 92 pages) and the Global Report (2009, 570 pages) are referred to in the Endnotes as IAAKSTD - Synthesis and IAAKSTD - Global.

Monsanto = a multinational corporation that is a world leader in GM organisms. This century-old empire has created some of the most toxic products ever sold, including polychlorinated-biphenyls (PCBs) and the herbicide Agent Orange. Today Monsanto has reinvented itself as a life sciences company converted to the virtues of sustainable development. Thanks to its genetically modified seeds, engineered among other things to withstand their herbicide Roundup, the world's best selling herbicide, the company claims it wants to solve world hunger while reducing environmental damage.

Nanotechnology = a technology that uses materials at an atomic or molecular scale, 1 to 100 nanometres, The technology has been used to create fast conductors for electronic uses, materials for solar cells, and materials used drugs and drug delivery systems, among others.

Oncomouse = a strain of inbred GM laboratory mouse created for cancer research by the addition or deletion of genes resulting in a strain of mouse that is more susceptible to cancer.

Precautionary principle = a principle used by public decision-makers that requires forbidding the introduction of a substance or process until it can be tested and deemed safe for the general public and the environment. This is in contrast to requiring proof of harm before a substance or process is banned. The precautionary principle is a statutory requirement in the law of the European Union.

Roundup = the brand name of abroad-spectrum herbicide produced by the U.S. company Monsanto, and contains the active ingredient glyphosate. Glyphosate is the most widely used herbicide in the U.S., and Roundup has been the number one selling herbicide worldwide since 1980. Several weed species, known as superweeds, have developed Roundup resistance largely because of repeated exposure.

Roundup Ready = plants that have been genetically engineered to be tolerant to glyphosate, the active ingredient in Monsanto's herbicide Roundup. The genes contained in these seeds are patented. Such crops allow farmers to use glyphosate as a post-emergence herbicide against most broadleaf and cereal weeds. Soy beans were the first Roundup Ready crop.

Stacked traits = inserting more than one gene for enhanced impact of a single benefit or for conferring multiple benefits, e.g., pest resistance and herbicide tolerance.

Terminator technology = a seed that is genetically engineered with a suicide mechanism that prevents the next generation of seeds from germinating the seed has had a terminator gene put into it. Since these seeds cannot germinate, seed saving becomes impossible. Traditional practice of seed saving is replaced by the need for farmers to buy seeds every time it is time to plant.

Transgenic organism = an organism into which a foreign gene has been introduced.

Endnotes: *(see Bibliography for full citation)*

1) Ursula Franklin, Professor of Physics Emeritus, University of Toronto Companion of the Order of Canada; Member of the Toronto Monthly Meeting of the Religious Society of Friends (Quakers); *The Real World of Technology* (1999)
2) Buntzel (2010)
3) A list of GM contents of commonly used food items in North America is found in Smith(2007), pp 258-9
4) Glaeser (1987)
5) Egger and Glaeser (1984); Ignacy Sachs quoted in Glaeser (1987)
6) Lathem (2009)
7) Roy (1987)
8) IFPRI (2010)
9) Berg (1975)
10) Fernandez-Cornejo and Caswell (2006)
11) Hammond (2011)
12) Benbrook (2002)
13) Pusztai and Bardocz (2011)
14) Pusztai and Bardocz (2011) p. 35
15) de Vendomois (2009)
16) Gurian-Sherman (2009)
17) IAAKSTD (2009)
18) Reservations about the 2009 IAAKSTD report from Australia, Canada, and the U.S. are published in IAAKSTD (2009) *Synthesis* Annex A and *Global* Annex G.
19) IAAKSTD (2009) p.391
20) Howard (1943)
21) Robin (2008)
22) Rodrigues (2009)
23) Gurian-Sherman (2009)
24) IFOAM (2010)
25) Kristiansen and Taji (2006)

26) EED (2010)
27) EED (2010) p. vii, and p. 114
28) Brown and Garver (2009)
29) Sale (1996)
30) Dawkins (2002), McHughen (2000), Charles (2001), Paarlberg (2001), and Shiva (2002)
31) Sen (1982)
32) Sharma (2004)
33) Advice from Dr. Tewolde Egziabher, Head of Ethiopia's Environmental Protection Agency, quoted in Pasternak, S., 2004
34) *Frontline* magazine (2010)
35) Wallace (2000)
36) Herring (2007)
37) Thies and Devare (2007) quoted in Herring (2007)
38) *Organizer* (2009) quoted in *Frontline* magazine (2010)
39) Brunk, Haworth, and Lee (1991)
40) Salter (1988)
41) Salter (1988), p. 181
42) Thurtle (2007)
43) Pillar and Yamamoto (1988)
44) Wheelis and Dando (2002)
45) Foley (2006)
46) Nelson (2001)
47) Brown and Garver (2009)
48) Gibson (2008)
49) Brown and Garver (2009)
50) Tansey and Rajotte (2008) pp. 214-215
51) Clarke (1987)
52) Ho (2010)

Bibliography *(websites accessed May 21, 2011)*

Benbrook, Charles, 2009. *Impacts of Genetically Engineered Crops on Pesticide Use: The First Thirteen Years.* Boulder, CO: The Organic Center. <organic-center.org/science.pest.php?action=view&report_id=159>

Brunk, Conrad G., Lawrence Haworth, and Brenda Lee, 1991. *Value Assumptions in Risk Assessment.* Waterloo: Wilfrid Laurier University Press.

Buntzel, Rudolf, 2010. International Lobbying on the Cartagena Protocol. EED, 2010 pp. 66 - 72.

Catt, Georgia, 2011. On farming and food. *The Economist*, March 19, 2011, p. 24.

Chennai, *Frontline*, PBS, March 12, 2010. *There was no listing on the Frontline for March 12, 2010, and Chennai seems to be a reporter who is on many stories.*

Charles, Daniel, 2001. *Lords of the Harvest: Biotech, Big Money, and the Future of Food.* New York: Perseus.

Clarke, George ed., 1993. *John Bellers, 1654 to 1725; Quaker Visionary; His Life, Times and Writings.* York, England: Sessions Book Trust.

Critical Art Ensemble, 2002. *The Molecular Invasion.* Brooklyn NY: Autonomedia.

Cummings, Claire Hope, 2008. *Uncertain Peril: Genetic Engineering and the Future of Seeds.* Boston: Beacon Press.

Dawkins, Kristin, 1997. *Gene Wars: The Politics of Biotechnology.* New York: Seven Stories Press.

Desantis, S'ra, 2004. Control through Contamination: Genetically Engineered Corn and Free Trade in Latin America. In *Gene Traders: Biotechnology, World Trade, and the Globalization of Hunger,* Brian Tokar, ed., Burlington, Vermont: Toward Freedom.

EED (Evangelische Entwicklungsdienst) and Partners, 2009. *Genetic Engineering and Food Sovereignty, Sustainable Agriculture is the Only Option to Feed the World.* Bonn, Germany Church Development Service EED. <eed.de/en/en.eed/en.eed.pub/en.pub.de.356/index.html>

Egger, K. and Glaeser, B., 1984. Kritik der Gruenen Revolution: Weg zur oekologischen Alternative In *Okologischer Landhau in den Tropen.* ed. P. Rottach, Karlsruhe, 1986, 2nd edition).

Fernandez-Cornejo, Jorge and Margriet Caswell, 2006. *The First Decade of Genetically Engineered Crops in the United States.* Washington: U. S. Department of Agriculture.

Foley, Duncan K., 2006. *Adam's Fallacy: A Guide to Economic Theology.* Cambridge MA: Harvard University Press.

Franklin, Ursula, 1999. *The Real World of Technology*, revised edition. Toronto: Anansi

Frontline magazine, Chennai, India, FVol. 27, No. 5 (Feb. 27 – March 12, 2010)

Gibson, Daniel G., *et al.*, 2008 Complete Chemical Synthesis, Assembly, and Cloning of a *Mycoplasma genitalium* Genome. *Science* 29 February 2008: 1215-1220.

Glaeser, Bernhard, 1987. Agriculture between the Green Revolution and Eco-development: Which Way to Go? In *The Green Revolution Revisited: Critique and Alternatives,* Bernhard Glaeser, ed. London: Unwin Hyman.

Gurian-Sherman, Doug, 2009. *Failure to Yield: Evaluating the Performance of Genetically Engineered Crops.* Cambridge MA: Union of Concerned Scientists. <ucsusa.org/assets/documents/food_and_agriculture/failure-to-yield.pdf>

Hammond, Edward, 2011. Genetically Engineered Backslide: The Impact of Glyphosate-Resistant Palmer Pigweed on Agriculture in the United States. *TWN Biotechnology & Biosafety Series* No. 12, Third World Network. <twnside.org.sg/title2/biosafety/pdf/bio12.pdf>

Herring, R. J., 2007 The Genomics Revolution and Development Studies: Science, Poverty and Politics. In *Transgenics and the Poor; Biotechnology in Development Studies,* Ronald Herring, *ed.* London: Routledge.

Ho, Mae-Wan, 2010. Golden Rice - An Exercise in How Not Do Science. *TWN Biotechnology and Biosafety Series* No.6 Third World Network. <twnside.org.sg/title/rice2.htm>

Howard, Sir Albert, 1943. *An Agricultural Testament.* London: Oxford University Press. <journeytoforever.org/farm_library/howardAT/ATtoc.html>

International Assessment of Agricultural Knowledge, Science and Technology for Development (IAAKSTD), 2009. *Agriculture at a Crossroads.* Washington: Island Press. <agassessment.org/reports>

International Food Policy Research Institute, 2010. *2010 Global Hunger Index.* Washington DC: International Food Policy Research Institute. <ifpri.org/publication/2010-global-hunger-index>

International Federation of Organic Agricultural Movements, 2010. *The World of Organic Agriculture.* Bonn, Germany: Research Institute of Organic Agriculture and International Federation of Organic Agricultural Movement. <organic-world.net/yearbook-2010.html>

Institute of Science and Society, 2010. GM Crops Facing Meltdown in the USA. *ISIS Report* . <i-sis.org.uk/GMCropsFacingMeltdown.php>

Khor, Martin, 2003. *Intellectual Property, Biodiversity and Sustainable Development.* London and New York: Zed Books.

Kristiansen, P., Acram Taji and John Regenhold, 2006. *Organic Agriculture: A Global Perspective.* Collingwood, Australia: CSIRO Publishing.

Lathem, Alexis, 2009. Upping the Stakes: Assessing the legacy of Norman Borlaug. Community College of Vermont Public Lecture. <vtcommons.org/blog/2009/10/08/upping-stakes-alexis-lathem-assessing-legacy-norman-borlaug>

Lappe, Francis Moore, Joseph Collins, Peter Rosset, and Luis Esparza, 1998. *World Hunger: Twelve Myths,* 2nd edition. New York: Grove Press.

Lappe, Anna, 2010. *Diet for a Hot Planet: The Climate Crisis at the End of Your Fork and What You can Do About It.* New York: Bloomsbury.

Lipton, Michael, 2007. Plant Breeding and Poverty: Can Transgenic Seeds Replicate the 'Green Revolution' as a Source of Gains for the Poor? In *Transgenics and the Poor;*

Biotechnology in Development Studies, Ronald Herring, ed. London: Routledge.

McHughen, Alan, 2000. *Pandora's Picnic Basket: The Potential and Hazards of Genetically Modified Foods.* New York: Oxford University Press.

Nelson, Robert H., 2001. *Economics as Religion: From Samuelson to Chicago and Beyond*, Chicago: University Park, PA: Pennsylvania State University Press.

Paarlberg, Robert, 2001. *Politics of Precaution.* Washington DC: International Food Policy Research Institute.

Pasternak, Shiri, 2004. Beggars Can't Be Choosers: GE Food Aid and the Threat of Food Sovereignty. In *Gene Traders: Biotechnology, World Trade, and the Globalization of Hunger,* Tokar, Brian, ed. Burlington, Vermont: Toward Freedom.

Patel, Raj, 2008. *Stuffed and Starved: The Hidden Battle for the World Food System.* London: Portobello.

Pillar, Charles and Keith Yamamoto, 1988. *Gene Wars: Military Control over the New Genetic Technologies.* New York: Beech Tree Books.

Pray, Carl E., Anwar Naseem, 2007. Supplying Crop Biotechnology to the Poor: Opportunities and Constraints. In *Transgenics and the Poor; Biotechnology in Development Studies, Ronald Herring, ed.* London: Routledge.

Pusztai, A and S. Bardocz, 2011. *Potential Health Effects of Foods Derived from Genetically Modified Plants*: *What are the Issues? TWN Biotechnology and Biosafety Series* No.14 Third World Network. <twnside.org.sg/title2/biosafety/pdf/bio14.pdf>

Robin, Marie-Monique, 2008. *The World According to Monsanto* (108 minute documentary film), Ontario, Canada: National Film Board of Canada.

Rodale, Maria, 2010. *Organic Manifesto: How Organic Farming Can Heal Our Planet, Feed the World, and Keep Us Safe.* New York: Rodale Books.

Rodrigues, Aruna, 2009. *Response to the FAO: How to Feed the World in 2050.* <dissidentvoice.org/2009/10/response-to-the-fao-how-to-feed-the-world-in-2050>

Roy, Rathindra Nath, 1987. Trees: appropriate tools for water and soil management. In *The Green Revolution Revisited: Critique and Alternatives,* Bernhard Glaeser, ed. London: Unwin Hyman.

Sahai, Suman, 2002. *Sowing Disaster's Seeds* <outlookindia.com/article.aspx?215861>

Sale, Kirkpatrick, 1996. *Rebels Against the Future: The Luddites and Their War Against the Industrial Revoluton.* New York: Basic Books

Salter, Liora, W. Leiss, and Edwin Levy, 1988. *Mandated Science: Science and Scientists in the Making of Standards.* New York: Springer

Sen, Amartya, 1983. *Poverty and Famines: An Essay on Entitlements and Deprivation.* London: Oxford University Press.

Sharma, Devinder, 2004. The Great Trade Robbery: World Hunger and the Myths of Industrial Agriculture. In *Gene Traders: Biotechnology, World Trade, and the Globalization of Hunger,* Tokar, Brian, ed. Burlington, Vermont: Toward Freedom.

Shiva, Vandana, 1999. *Biopiracy: The Plunder of Nature and Knowledge.* Cambridge MA: South End Press.

Shiva, Vandana, 2002. *Protect or Plunder? Understanding Intellectual Property Rights.* London and New York: Zed Books.

Shiva, Vandana, ed., 2007. *Manifestos on the Future of Food and Seed.* Cambridge MA: South End Press.

Smith, Jeffrey M., 2003. *Seeds of Deception: Exposing Industry and Government Lies About the Safety of the Genetically Engineered Foods You're Eating.* Fairfield, Iowa: Yes! Books.

Smith, Jeffrey M., 2007. *Genetic Roulette: The Documented Health Risks of Genetically Engineered Foods.* Fairfield, Iowa: Yes! Books.

Soil Association, 2002. *Seeds of Doubt: North American Farmers' Experience of GM Crops.* Briston, England: Soil Association. <grain.org/docs/seeds-of-doubt.pdf>

Tansey, Geoff and Tasmin Rajotte, eds., 2008. *The Future Control of Food: A Guide to International Negotiations and Rules on Intellectual Property Rights, Biodiversity and Food*

Security. (Quaker International Affairs Programme, Ottawa and International Development Research Centre, Ottawa) London/Sterling VA: Earthscan. <idl-bnc.idrc.ca/dspace/bitstream/10625/35059/1/127132.pdf>

Thies. Janice E., and Medha H. Devare, 2007. An Ecological Assessment of Transgenic Crops. In *Transgenics and the Poor; Biotechnology in Development Studies.*, Ronald Herring, ed. London: Routledge.

Thurtle, Phillip, 2008. *The Emergence of Genetic Rationality.* Seattle WA: University of Washington Press.

Tokar, Brian, ed., 2004. *Gene Traders: Biotechnology, World Trade, and the Globalization of Hunger.* Burlington, Vermont: Toward Freedom.

Tudge, Colin, 2007. *Feeding People Is Easy.* Grosseto, Italy: Pari Publishing.

Vandana, V. A. H. Jafri, A. Emani and M. Pande, 2002. *Seeds of Suicide: Ecological and Human Costs of Globalization of Agriculture.* New Delhi, India: Research Foundation for Science, Technology, and Ecology.

de Vendomois Joël Spiroux, François Roullier, Dominique Cellier, Gilles-Eric Séralini., 2009. A Comparison of the Effects of Three GM Corn Varieties on Mammalian Health. *International Journal of Biological Sciences* 5: 706 – 726.

Wallace, Helen, 2000. Health, agriculture and the development of the 'knowledge-based economy'. In *Policy-Making in the European Union,* 4th edition, Helen Wallace and William Wallace, eds. Oxford: Oxford University Press.

Wheelis, Mark and Malcolm Dando, 2002. On the Brink: Biodefence, Biotechnology, and the Future of Weapons Control. *Chemical and Biological Weapons Control Bulletin*, December 2002.

World Neighbors, 1982. *Two Ears of Corn, A Guide to People-Centered Agricultural Improvement.* Oklahoma City, Oklahoma: World Neighbors.

Zilberman, David, Holly Ameden and Matin Qaim, 2007. The Impact of Agricultural Biotechnology on Yields, Risks, and Biodiversity in Low-Income Countries. In *Transgenics and the Poor; Biotechnology in Development Studies.*, Ronald Herring, ed. London: Routledge.

Contributors

Anne Mitchell is a member of Toronto Monthly Meeting of the Religious Society of Friends (Quakers). She is the General Secretary of Quaker Earthcare Witness. Anne is Clerk of the Canadian Yearly Meeting—the national body of Quakers in Canada. Anne has served on various boards, committees, advisory bodies and multi-stakeholder processes, and currently serves on the Board of Directors of the Quaker Institute for the Future, the Quaker International Affairs Program and on the Biotechnology Reference Group of the Canadian Council of Churches. Anne attended the World Council of Churches Global Consultation on Genetics and New Biotechnologies in 2006.

Anne's particular interest is to see policy change that supports the public good and sustainability, and to see how Quaker discernment processes can be applied to these questions. Anne was Executive Director of the Canadian Institute for Environmental Law and Policy (CIELAP) from 1992–2009 and remains CIELAP's a Senior Advisor on sustainability. Anne has led other non-profit organizations, with a focus on community development and human rights, at the international and national levels. She has also been a high school teacher and group counsellor. She has a Master of Arts degree in International Affairs from the Norman Paterson School of International Affairs at Carleton University, Ottawa. Anne is a recognized leader and change agent.

Pinayur Rajagopal is a member of the Toronto Monthly Meeting of the Religious Society of Friends. He has served on several Monthly Meeting bodies including the Finance and Refugee Committees, and Meeting of Ministry and Counsel. He has been a member and clerk of the Canadian Friends Service Committee when it was a standing committee of the Canadian Yearly Meeting.

Raja has been a student of Mathematics all his life; he obtained a Master's degree from Madras University (India) and Ph. D. from Cambridge (England). While at Cambridge, he found the new science of computing to be an alluring associate of Mathematics, and has worked in both areas ever since. He worked at Max Planck Institute in Munich before coming to York University in Toronto. He has taught students at all levels in the University. He has published in the areas of his research specialties, including articles on the transmission of Indian Mathematics to the West via Baghdad, Damascus and Venice. He has coauthored papers on the changes in the financing of Universities and other institutions of higher learning and research in the last part of the 20th century. These publications have focused on the way research results have changed from being regarded as societal property to becoming the property of corporations that entitles them to patents and marketable products. Raja has a longstanding interest in sustainable food systems and undertook his research survey on what scientific journals were reporting on genetically modified crops at the request of Ann Mitchell for this QIF Pamphlet project.

Keith Helmuth is a member of New Brunswick Monthly Meeting of the Religious Society of Friends and a founding Board Member of Quaker Institute for the Future (QIF). He taught environmental studies and social ecology at Friends World College in New York and economic development at the College's East African Centre in Kenya. In 1971 he and his wife, Ellen established North Hill Farm as a family business in Carleton County, New Brunswick, Canada, with a particular interest in local food system development. Keith has been a community economic development activist working with producer and marketing cooperatives in his home region. He helped establish and develop the Woodstock Farm Market Cooperative and served on the Board of his local credit union for twenty-five years.

He is a contributing author to, *Right Relationship: Building a Whole Earth Economy* (2009), and *How on Earth Do We Live Now? Natural Capital, Deep Ecology, and the Commons* (2011). He recently published a study of John Bellers (1654 – 1725), *The Evolutionary Potential of Quakerism Revisited* (2011). Keith is coordinator of QIF's Circles of Discernment programme and a member of its Publications Committee. After retiring from farming Keith spent ten years as Manager of Penn Book Center on the campus of the University of Pennsylvania in Philadelphia where he also worked with Quaker projects on the conflict between economics and ecology. Now back in New Brunswick, he is again on the Board of the Woodstock Farm Market, and a member of Transition Town Woodstock and the Woodstock Sustainable Energy Group.

Susan Holtz Susan Holtz has been an energy and environmental policy analyst for some 35 years, most of that time in Nova Scotia, where she was a long-time member of Halifax Monthly Meeting of the Religious Society of Friends. From 2000 to 2009 she lived in Toronto, where among other things she did projects for the Canadian Institute for Environmental Law and Policy on policy issues involving biotechnology, nanotechnology, and pharmaceutical and personal care product contamination of water resources. Her latest publications included several chapters of the 2009 Earthscan book, *Making the Most of the Water We Have: The Soft Path Approach to Water Management.*

Recently retired, she is an active member of the County Sustainability Group in Prince Edward County, Ontario, Canada, where she and her husband Fred now live. In 2010 Susan and Fred transferred their Quaker membership to Wooler Monthly Meeting of Wooler, Ontario. Susan was the founding Vice Chair of both the Canadian and Nova Scotia Round Tables on the Environment and the Economy. She was also instrumental in setting up the Canadian Environmental Network. She has served on many advisory bodies and boards, including the Auditor General of Canada's Panel of Senior Advisors and the Canadian Commissioner of Environment and Sustainable Development's External Advisory Panel. She has authored many papers and reports on public policy and environmental issues.

Quaker Institute for the Future

The Quaker Institute for the Future (QIF) seeks to generate systematic insight, knowledge, and wisdom that can inform public policy and enable us to treat all humans, all communities of life, and the whole Earth as manifestations of the Divine. QIF creates the opportunity for Quaker scholars and practitioners to apply the social and ecological intelligence of their disciplines within the context of Friends' testimonies and the Quaker traditions of truth seeking and public service.

The focus of the Institute's concerns include:

- Economic behavior that increasingly undermines the ecological processes on which life depends.

- The development of technologies and capabilities that hold us responsible for the future of humanity and the Earth.

- Structural violence and lethal conflict arising from the pressures of change, increasing inequity, concentrations of power and wealth, declining natural capital, and increasing militarism.

- The increasing separation of people into areas of poverty and wealth, and into social domains of aggrandizement and deprivation.

- The philosophy of individualism and its socially corrosive promotion as the principal means for the achievement of the common good.

- The complexity of global interdependence and its demands on governance systems and citizen's responsibilities.

- The convergence of ecological and economic breakdown into societal disintegration.

QIF Board of Trustees: Gray Cox, Elaine Emmi, Phil Emmi, Geoff Garver, Leonard Joy, Keith Helmuth, Laura Holliday, Anne Mitchell, and Shelley Tanenbaum.

Quaker Institute for the Future
<quakerinstitute.org>

www.ingramcontent.com/pod-product-compliance
Lightning Source LLC
Chambersburg PA
CBHW032214040426
42449CB00005B/588